A REVIEW OF GAO'S FINDINGS AND RECOMMENDATIONS REGARDING THE DEPARTMENT OF ENERGY'S EFFORTS TO CONSOLIDATE SURPLUS PLUTONIUM INVENTORIES

HEARING

BEFORE THE

SUBCOMMITTEE ON OVERSIGHT AND INVESTIGATIONS

OF THE

COMMITTEE ON ENERGY AND COMMERCE

HOUSE OF REPRESENTATIVES

ONE HUNDRED NINTH CONGRESS

FIRST SESSION

OCTOBER 7, 2005

Serial No. 109–56

Printed for the use of the Committee on Energy and Commerce

Available via the World Wide Web: http://www.access.gpo.gov/congress/house

U.S. GOVERNMENT PRINTING OFFICE

24–254PDF WASHINGTON : 2005

For sale by the Superintendent of Documents, U.S. Government Printing Office
Internet: bookstore.gpo.gov   Phone: toll free (866) 512–1800; DC area (202) 512–1800
Fax: (202) 512–2250   Mail: Stop SSOP, Washington, DC 20402–0001

COMMITTEE ON ENERGY AND COMMERCE

JOE BARTON, Texas, *Chairman*

RALPH M. HALL, Texas
MICHAEL BILIRAKIS, Florida
  Vice Chairman
FRED UPTON, Michigan
CLIFF STEARNS, Florida
PAUL E. GILLMOR, Ohio
NATHAN DEAL, Georgia
ED WHITFIELD, Kentucky
CHARLIE NORWOOD, Georgia
BARBARA CUBIN, Wyoming
JOHN SHIMKUS, Illinois
HEATHER WILSON, New Mexico
JOHN B. SHADEGG, Arizona
CHARLES W. "CHIP" PICKERING,
  Mississippi, *Vice Chairman*
VITO FOSSELLA, New York
ROY BLUNT, Missouri
STEVE BUYER, Indiana
GEORGE RADANOVICH, California
CHARLES F. BASS, New Hampshire
JOSEPH R. PITTS, Pennsylvania
MARY BONO, California
GREG WALDEN, Oregon
LEE TERRY, Nebraska
MIKE FERGUSON, New Jersey
MIKE ROGERS, Michigan
C.L. "BUTCH" OTTER, Idaho
SUE MYRICK, North Carolina
JOHN SULLIVAN, Oklahoma
TIM MURPHY, Pennsylvania
MICHAEL C. BURGESS, Texas
MARSHA BLACKBURN, Tennessee

JOHN D. DINGELL, Michigan
  Ranking Member
HENRY A. WAXMAN, California
EDWARD J. MARKEY, Massachusetts
RICK BOUCHER, Virginia
EDOLPHUS TOWNS, New York
FRANK PALLONE, Jr., New Jersey
SHERROD BROWN, Ohio
BART GORDON, Tennessee
BOBBY L. RUSH, Illinois
ANNA G. ESHOO, California
BART STUPAK, Michigan
ELIOT L. ENGEL, New York
ALBERT R. WYNN, Maryland
GENE GREEN, Texas
TED STRICKLAND, Ohio
DIANA DeGETTE, Colorado
LOIS CAPPS, California
MIKE DOYLE, Pennsylvania
TOM ALLEN, Maine
JIM DAVIS, Florida
JAN SCHAKOWSKY, Illinois
HILDA L. SOLIS, California
CHARLES A. GONZALEZ, Texas
JAY INSLEE, Washington
TAMMY BALDWIN, Wisconsin
MIKE ROSS, Arkansas

BUD ALBRIGHT, *Staff Director*
DAVID CAVICKE, *Deputy Staff Director and General Counsel*
REID P.F. STUNTZ, *Minority Staff Director and Chief Counsel*

———

SUBCOMMITTEE ON OVERSIGHT AND INVESTIGATIONS

ED WHITFIELD, Kentucky, *Chairman*

CLIFF STEARNS, Florida
CHARLES W. "CHIP" PICKERING,
  Mississippi
CHARLES F. BASS, New Hampshire
GREG WALDEN, Oregon
MIKE FERGUSON, New Jersey
MICHAEL C. BURGESS, Texas
MARSHA BLACKBURN, Tennessee
JOE BARTON, Texas,
  (Ex Officio)

BART STUPAK, Michigan
  Ranking Member
DIANA DeGETTE, Colorado
JAN SCHAKOWSKY, Illinois
JAY INSLEE, Washington
TAMMY BALDWIN, Wisconsin
HENRY A. WAXMAN, California
JOHN D. DINGELL, Michigan,
  (Ex Officio)

CONTENTS

(III)

A REVIEW OF GAO'S FINDINGS AND RECOMMENDATIONS REGARDING THE DEPARTMENT OF ENERGY'S EFFORTS TO CONSOLIDATE SURPLUS PLUTONIUM INVENTORIES

FRIDAY, OCTOBER 7, 2005

House of Representatives,
Committee on Energy and Commerce,
Subcommittee on Oversight and Investigations,
Washington, DC.

The subcommittee met, pursuant to notice, at 9 a.m., in room 2123, Rayburn House Office Building, Hon. Ed Whitfield (chairman) presiding.

Members present: Representatives Whitfield, Burgess, Blackburn, Stupak, and Inslee.

Staff present: Dwight Cates, majority professional staff; Mark Paoletta, chief counsel; Terry Lane; deputy communications director; Jonathan Pettibon, clerk; and Chris Knauer, minority professional staff.

Mr. WHITFIELD. I will call the meeting to order, and the subject of today's hearing is a review of the GAO's findings and recommendations regarding the Department of Energy's efforts to consolidate surplus plutonium inventories.

I certainly want to thank our witnesses for being here today, Mr. Gene Aloise, who is the Director of the Natural Resources and the Environment, U.S. Government Accountability Office; Mr. Charles Anderson, who is the Principal Deputy Assistant Secretary of Environmental Management at the United States Department of Energy, and Dr. A.J. Eggenberger, who is Chairman of the Defense Nuclear Facility Safety Board.

We appreciate all of you being here and we certainly look forward to your testimony. At this time I would give my opening statement.

Today the subcommittee will review the findings and recommendations of a Government Accounting Office report requested by this committee regarding the Department of Energy's efforts to consolidate plutonium inventories. Over the past several years the Oversight and Investigations Subcommittee has held several hearings on the status of security at DOE nuclear facilities.

Several DOE sites contain tons of nuclear material that could be used against the U.S. if they fell into the hands of a terrorist organization. Consequently the security and protection of domestic DOE nuclear facilities has been a critical first line of defense against terrorism.

The Department has made several significant changes to its security policy in the years since September 11th, 2001, that have resulted in considerable improvements to physical security at each DOE nuclear facility. These improvements are largely driven by changes to the Design Basis Threat, a classified document that estimates the size and characteristicsof an advisory force that each nuclear facility must defend against. In response, the Department made immediate upgrades to physical security and hired more guards in the aftermath of 9/11.

The Department continues to refine the characteristics of the postulated adversary, and the Design Basis Threat was most recently revised in October of 2004. Each change to the Design Basis Threat has required additional security upgrades and more guards.

In our May 2004 hearing, then Deputy Secretary Kyle McSlarrow identified several additional security initiatives the Department would undertake. Among these, Deputy Secretary McSlarrow outlined an important initiative to consolidate and centralize the storage of high-risk nuclear materials. The consolidation of nuclear materials stored at multiple facilities down to just a few facilities could result in security improvements and significant cost savings.

I am please that Secretary Bodman has committed to continue the Department's initiative to consolidate these materials. Earlier this year the Secretary created a Nuclear Materials Consolidation and Coordination Committee to determine how and where a nuclear materials facility should be consolidated.

Today's hearing will focus on the findings and recommendations of the GAO that reviewed the Department's ongoing efforts to consolidate surplus plutonium inventories at the Savannah River site in South Carolina.

Mr. Gene Aloise, Director of Natural Resources and Environment, will provide the testimony about GAO's findings and recommendations. GAO's written testimony indicates that the Department has had considerable trouble over the past several years following through on its commitment to consolidate plutonium inventories at Savannah River.

Some of these difficulties relate to technical issues that are not easily resolved, while other problems stem from the lack of a comprehensive strategy and ineffective coordination among the many sites that store the plutonium.

We strongly endorse GAO's recommendations that DOE develop a comprehensive strategy to consolidate, store, and eventually dispose of its plutonium. I hope the Department can soon finalize a disposition plan for the plutonium and complete the necessary safety upgrades at Building 105-K at the Savannah River site so that we can move forward with plutonium storage and consolidation.

I hope the Department will continue to investigate all available consolidation options, including opportunities to consolidate nuclear materials at other DOE sites where existing buildings can be used for nuclear materials storage.

We look forward to hearing the testimony of Mr. Charles Anderson as well, and certainly Dr. Eggenberger. And with that I would recognize the ranking minority member, Mr. Stupak.

[The prepared statement of Hon. Ed Whitfield follows:]

PREPARED STATEMENT OF HON. ED WHITFIELD, CHAIRMAN, SUBCOMMITTEE ON OVERSIGHT AND INVESTIGATIONS

This hearing will come to order. Today the Subcommittee will review the findings and recommendations of a Government Accountability Office report, requested by this committee, regarding the Department of Energy's efforts to consolidate plutonium inventories.

Over the past several years, the Oversight and Investigations Subcommittee has held numerous hearings on the status of security at DOE nuclear facilities. Several DOE sites contain tons of nuclear material that could be used in devastating attacks if they fell into the hands of a terrorist organization. Consequently, the security and protection of domestic DOE nuclear facilities has been a critical first line of defense against terrorism.

The Department has made several significant changes to its security policy in the years since September 11th , 2001, that have resulted in considerable improvements to physical security at each DOE nuclear facility. These improvements are largely driven by changes to the "design basis threat"—a classified document that estimates the size and characteristics of an adversary force that each nuclear facility must defend against. In response, the Department made immediate upgrades to physical security and hired more guards in the aftermath of 9-11. The Department continues to refine the characteristics of the postulated adversary, and the design basis threat was most recently revised in October of 2004. Each change to the design basis threat has required additional security upgrades and more guards.

At our May 2004 hearing, then-Deputy Secretary Kyle McSlarrow identified several additional security initiatives the Department would undertake. Among these, Deputy Secretary McSlarrow outlined an important initiative to consolidate and centralize the storage of high-risk nuclear materials. The consolidation of nuclear materials stored at multiple facilities down to just a few facilities could result in security improvements and significant cost savings.

I am pleased that Secretary Bodman is committed to continue the Department's initiative to consolidate nuclear materials. Earlier this year Secretary Bodman created a Nuclear Materials Consolidation and Coordination Committee to determine how and where nuclear materials should be consolidated.

Today's hearing will focus on the findings and recommendations of the GAO that reviewed the Department's ongoing efforts to consolidate surplus plutonium inventories at the Savannah River Site in South Carolina. Mr. Gene Aloise, Director of Natural Resources and Environment, will provide testimony about GAO's findings and recommendations.

GAO's written testimony indicates that the Department has had considerable trouble over the past several years following-through on its commitment to consolidate plutonium inventories at the Savannah River Site. Some of these difficulties relate to technical issues that are not easily resolved, while other problems stem from the lack of a comprehensive strategy and ineffective coordination among the many sites that store plutonium.

I strongly endorse GAO's recommendation that DOE "develop a comprehensive strategy to consolidate, store, and eventually dispose of its plutonium."

I hope the Department can soon finalize a disposition plan for plutonium, and complete the necessary safety upgrades at building 105-K at the Savannah River Site so that we can move forward with plutonium storage and consolidation. I hope the Department will continue to investigate all available consolidation options, including opportunities to consolidate nuclear materials at other DOE sites where existing buildings could be used for nuclear material storage.

I look forward to hearing the testimony of Mr. Charlie Anderson, Principal Deputy Assistant Secretary for Environmental Management, regarding DOE's plans for plutonium consolidation, as well as the testimony of Dr. A.J. Eggenberger of the Defense Nuclear Facility Safety Board. I yield back.

Mr. STUPAK. Thank you, Mr. Chairman. Mr. Chairman, today is a hearing on plutonium consolidation and the findings by GAO that DOE has still not made adequate progress in this area.

While this is an important subject, we were only given notice of this hearing late last Friday, and unfortunately we have a major energy bill on the floor today. I and many others from this committee will need to be present for that debate, so we will do our best to be in both places.

Mr. Chairman, I do want to thank GAO for the report that they provided to this committee. They will discuss today that DOE still cannot consolidate its excessive plutonium at the Savannah River site, despite having been on notice to do so for several years.

As GAO has already noted in the report issued in July, DOE was unable to complete a plan to process the plutonium into a form for permanent storage. This is a requirement under the National Defense Authorization Act. Unfortunately, because no plan was in place, DOE is not allowed to ship any additional plutonium to Savannah River until such a plan is created.

Also, according to GAO, Savannah River cannot receive all of the plutonium from DOE's Hanford site because it is not in a form that Savannah River planned to receive and store.

Mr. Chairman, GAO will also tell us that DOE lacks the capability to fully monitor the condition of the plutonium necessary to ensure continual safe storage. According to GAO, the facility at Savannah River that DOE intends to use for material storage purposes lacks adequate safety systems, including proper monitoring capability. Without proper monitoring capability, DOE faces increased risk of an accidental plutonium release that could harm the public, workers and the environment.

Mr. Chairman, I know you agree with me that this is unacceptable, and I would urge you to have this matter addressed and even bring the Secretary of Energy before the committee to discuss the matter if we wish to see faster results.

If we do not hold these agencies and departments accountable for the repeated problems that GAO and the Inspectors General keep finding, then the issue will never beresolved, at least not in our lifetime.

Mr. Chairman, let me conclude by saying that I notice that Secretary Bodman was invited to this hearing but declined to attend. This is unfortunate, because had he been here I would have had a number of questions for him about the high fuel prices we are expected to see again later this year.

As you know, I have repeatedly requested that this committee hold hearings on the skyrocketing cost of home heating fuel oil, including oil, propane and natural gas. My constituents live in the coldest districts of this country. In fact last night it was in the mid-20's and we are having snow.

They are being told that in some cases they should expect their energy bills to more than double this winter. For some this may mean the difference between staying warm or staying fed. Incredibly the Secretary of Energy's solution to this is to insulate our houses, drive less, and change our light bulbs. Trust me, my constituents know how to do these common sense things and have done them for years.

But more importantly, they would like to know why energy prices are expected to more than double this year. They would also like to know that they are not being gouged by unfair industry practices.

While the Secretary of Energy did note that much of the expected increaseswere due to Hurricane Katrina and Rita, he also said that he did not know the extent of the damage caused by these two storms. How then does he know that the expected price increases in energy are justified? How do we know that our constituents are not being gouged?

The Secretary of Energy does not have a full understanding of the condition of the gulf coast energy infrastructure and neither do we on this committee. I would note for the record that the oversight committee has not asked for nor received a single briefing on the state of the energy infrastructure in the gulf region since the Hurricane. We do not know what is damaged, how much is damaged, and what impacts such destruction should have on energy prices.

As the investigative arm of the Energy and Commerce Committee, I find no compelling reason why we choose to keep our head on such an important issue. I also find no compelling reason why we do not choose to investigate why energy prices are set to climb through the roof and why we are not allowed to determine whether these price increases are justified.

As the Energy and Commerce Committee, we owe this to the American public. At least, I owe it to my constituents, who are expecting a very cold winter with very high fuel bills.

I yield back the balance of my time, and thank you, Mr. Chairman.

Mr. WHITFIELD. Thank you, Mr. Stupak. At this time I recognize Mr. Burgess for his opening statement.

Mr. BURGESS. Thank you, Mr. Chairman. I think in the interests of time let me submit my opening statement for the record and we will hear from the witnesses.

[The prepared statement of Hon. Michael C. Burgess follows:]

PREPARED STATEMENT OF HON. MICHAEL C. BURGESS, A REPRESENTATIVE IN CONGRESS FROM THE STATE OF TEXAS

Thank you Mr. Chairman, and thank you for having this important hearing.

The issue of nuclear waste storage is an extremely timely issue and will only become more important as we look to nuclear power as an emissions-free source of electricity.

Congress has acknowledged that consolidating nuclear waste storage is advantageous because of lower costs and increased security. While it is my hope that we can soon move forward with plans for a permanent disposal facility, such as Yucca Mountain, we must have an interim plan in place until that facility can be constructed.

Today we will review a report from the Government Accountability Office that was originally commissioned nearly two years ago. The report examines the DOE's management practices as they have worked to consolidate excess plutonium inventories. I was concerned to read that a lack of coordination by DOE management may cost the federalgovernment an additional $85 million per year in order to continue to store plutonium at the Hanford site.

The security ramifications of leaving the plutonium inventories at Hanford are outside the purview of this report, but are troubling nonetheless. At this time in American history, our national security has become the most important issue facing our nation. It is critical that we do everything within our power to ensure that our nuclear weapons and nuclear materials are highly secured and protected.

I am pleased to see that the Department of Energy has concurred with the recommendations in the report and hope that they will move to implement them with alacrity. We cannot allow our national security to be compromised because of inadequate safeguards that are poorly implemented.

Again, Mr. Chairman, I thank you for this crucial hearing in which we can address some of these essential concerns regarding nuclear facilities and the security of our nation.

Mr. WHITFIELD. Okay. Well, that concludes the opening statements. As Mr. Stupak said, we do have an energy bill on the floor, and we don't expect our first votes until around 10:15, so although we may appear to be in a hurry this is such an important issue that we do want to take whatever time is necessary, and we look forward to your testimony.

And, Mr. Aloise, we will begin with you, although you are at the reverse end. But you will be our first witness. So you will be recognized for 5 minutes.

As you realize when we do have oversight and investigation hearings, we normally do swear the witnesses in. And I would ask you, do any of you have any difficulty with testifying under oath in morning?

As you realize, the rules of the House and rules of this committee are that you are entitled to legal counsel, if you so choose. Do any of you prefer to have legal counsel with you this morning?

In that case then if you would simply rise and raise your right hand.

[Witnesses sworn.]

Mr. WHITFIELD. You are now under oath. And Mr. Aloise, you may now give a 5 minute opening statement. Thank you.

TESTIMONY OF GENE ALOISE, DIRECTOR, NATURAL RESOURCES AND THE ENVIRONMENT, U.S. GOVERNMENT ACCOUNTABILITY OFFICE; CHARLES E. ANDERSON, PRINCIPAL DEPUTY ASSISTANT SECRETARY OF ENVIRONMENTAL MANAGEMENT, U.S. DEPARTMENT OF ENERGY; AND A.J. EGGENBERGER, CHAIRMAN, DEFENSE NUCLEAR FACILITY SAFETY BOARD; ACCOMPANIED BY JACK MANSFIELD AND BRUCE MATTHEWS

Mr. ALOISE. Thank you, Mr. Chairman. Mr. Chairman and members of the subcommittee, I am pleased to be here today to discuss our work on DOE's efforts to consolidate surplus plutonium.

DOE stores about 50 metric tons of plutonium that is no longer needed for nuclear weapons. Most of this surplus plutonium is pit form, and is stored at the Pantex plant in Texas. The remaining plutonium, primarily contaminated metals and oxides, is stored at several locations across the United States, including the Savannah River and Hanford sites, Los Alamos and Lawrence Livermore National Labs.

It is important that this plutonium be consolidated for health, safety and security reasons. DOE has not yet decided where to consolidate this plutonium, but has created enough storage space at the Savannah River site in the event the plutonium is consolidated there. Eventually, this plutonium must be processed in a form appropriate for permanent disposal.

My remarks, which are based on our July 2005 report, will focus on the extent to which DOE can consolidate plutonium at the Savannah River site and the site's capacity to monitor the safety of the plutonium storage containers.

Regarding consolidation, DOE cannot consolidate all of its plutonium at the Savannah River site because it has not completed a plan to process the plutonium into a form for permanent disposal. This plan was required by the National Defense Authorization Act of 2002. However, even if DOE were able to ship the plutonium, other problems stand in the way of consolidation.

Specifically, DOE approved Hanford's overall cleanup plan and Savannah River's plutonium storage plan, even though the plans are inconsistent with one another. Hanford's cleanup plan calls for shipping about 20 percent of the plutonium in the form of 12-foot long nuclear fuel rods. Savannah River's plan assumed that all of Hanford's plutonium would be shipped in DOE's standard 5-inch wide and 10-inch long storage containers.

Subsequently DOE determined that Savannah River had the space to store the fuel rods, but they cannot be shipped because, among other things, there is currently no Department of Transportation approved shipping container.

Wherever the fuel rods end up, they will have to be disassembled prior to processing them for permanent disposal, and neither Hanford nor Savannah River have the capability to disassemble them.

The challenges DOE faces storing its plutonium, in our view, stem from the Department's failure to adequately plan for plutonium consolidation. Instead of developing an integrated plan, DOE relied on its individual sites to independently develop plans to achieve their own goals.

As a result, DOE will not achieve the cost savings and security improvements that consolidation offers. In fact, continued storage of all of Hanford's plutonium will cost approximately $855 million more per year because of rising security costs.

Regarding the safe monitoring of plutonium, the Savannah River facility where DOE is storing the plutonium is not equipped to conduct the needed monitoring of the storage containers. That is, it lacks adequate fire protection, ventilation and filtration.

DOE planned to construct a monitoring capability in another building at the site which already had the ventilation systems needed to work with plutonium; however, this building would not have sufficient security to conduct all of the required monitoring and had other serious safety concerns as well.

Because of these concerns, DOE plans to install monitoring equipment and the necessary safety systems in the building where the plutonium is now stored. In our July report, we made recommendations to ensure that DOE develops a comprehensive strategy for plutonium consolidation, storage and disposal, and that its facilities cleanup plans are consistent with this strategy.

DOE agreed with our recommendations and said that it will develop a strategic plan for the consolidation of plutonium and other nuclear materials.

Mr. Chairman, that concludes my remarks. I would be happy to respond to questions you or members of the subcommittee may have.

Mr. WHITFIELD. Thank you very much for your testimony, and at this time, I recognize Dr. Charles Anderson.

TESTIMONY OF CHARLES E. ANDERSON

Mr. ANDERSON. Good morning. My name is Charles Anderson. I am the Department of Energy's Principal Deputy Assistant for the Office of Environmental Management. I have been involved with plutonium disposition for a number of years in this and previous positions. I am also a member of the Department's Nuclear Materials Disposition and Consolidation Coordination Committee. The

Office of Environmental Management is responsible for the safe storage and security of the majority of DOE's surplus non-pit plutonium pending disposition of that material.

The GAO's July 2005 report contains two recommendations for executive action with which the Department concurs. The first recommendation calls for a comprehensive strategy to be developed for the consolidation, storage and disposition of DOE's excess plutonium.

The second recommendation suggests DOE's cleanup plans be reviewed to ensure they are consistent with this comprehensive strategy for consolidation, storage and disposition.

While the Department believes that consolidation can result in significant benefits with respect to safety and cost savings, any future decisions to do so will be based on the outcome of ongoing evaluations and considerations that will provide the foundation for the development of the Nuclear Materials Disposition and Consolidation Coordination Committee's strategic plan.

The Department remains committed to conduct a disposition of plutonium and any further consolidation of the material in a manner that is good for the environment, safe for the worker, respectful of the taxpayer, and consistent with all applicable statutory requirements. Until the Nuclear Materials Disposition and Consolidation Coordination Committee's s strategic plan is completed, the Department will not make a decision on plutonium consolidation.

With respect to the first GAO recommendation that the Department develop a comprehensive strategy for consolidation, storage and disposition of the Department's excess plutonium, former Secretary Abraham established this committee. Secretary Bodman subsequently approved the charter for this committee. A key responsibility of this committee is to develop and ensure implementation of a strategic plan for disposition and consolidation of special nuclear material. This strategic plan will encompass the comprehensive strategy recommended by the GAO.

The principal mission of this committee is to provide the Department with recommendations on cross-cutting nuclear materials disposition and consolidation planning with the objectives of providing the necessary security for DOE's nuclear materials, identifying paths for disposition, as appropriate, and reducing out-year security and program costs. The scope of the material within the committee's charter includes all of the surplus plutonium owned by the office of environmental management, and also surplus non-pit plutonium owned by the National Nuclear Security Administration.

Deputy Secretary Sell recently approved the mission need for a plutonium disposition project at the Savannah River site for plutonium that does not have an identified disposition path; that is, plutonium not suitable for disposition using the currently designed mixed oxide, or MOX, fuel fabrication facility planned to be constructed at the Savannah River site. The Department's fiscal year 2006 Congressional budget request includes $10 million for conceptual design of the Savannah River site plutonium disposition project. As part of this conceptual design, the Department will be evaluating a number of alternatives to meet the disposition objective.

In response to the GAO's second recommendation, I can assure this Congressional committee that following the completion of the NMDCCC's strategic plan, the Office of Environmental Management and its staff will revise and review the appropriate cleanup plans to make certain they are consistent with this strategic plan and its associated implementation schedule.

I would like to say a few words about the findings contained in the GAO's report which are based on the premise that plutonium will be consolidated at the Savannah River site and have in no way been approved or endorsed by the Department.

First, prior to shipping any additional weapons usable plutonium to the Savannah River site, the Department will comply with all applicable statutory requirements, including those established by the National Defense Authorization Act for Fiscal Year 2002 and by the National Defense Authorization Act for Fiscal Year 2003.

With respect to being able to receive all of Hanford's plutonium, the Savannah River site's K-area currently has adequate storage capacity for all of Hanford's plutonium, including the unirradiated fuel rods now stored at Hanford. While additional activities need to be completed, including development of a revised safety documentation, shipping package certification, and appropriate National Environmental Policy Act, or NEPA, analysis, the Hanford fuel rods can be shipped and stored intact at the Savannah River site.

While recent changes to DOE's security requirements make it highly desirable to have as few nuclear materials storage locations as possible, the elimination of one facility planned for use of plutonium storage at the Savannah River site, the Metallurgical Building, building 235-F, does not complicate our potential storage plans since K-area alone now has adequate storage capacity. Furthermore, the Department agrees with the Defense Nuclear Facilities Safety Board and that there are potential safety issues associated with continued use of the Metallurgical Building.

Finally, with respect to the GAO finding that DOE lacks capability to fully monitor the condition of the plutonium necessary to ensure continued safe storage, the existing plutonium surveillance and monitoring capabilities in the Metallurgical Building are all that are required until 2007 to ensure continued safe storage. Beginning in 2007, the Savannah River site will have the capability in K-area to perform all required surveillance and monitoring examinations to ensure safe storage of plutonium at the site.

In closing, it is very important to keep in mind that while the Department is evaluating the options for the safe and secure storage of weapons usable plutonium, the Department currently has no plans nor have we made any decisions to further consolidate such plutonium to the Savannah River site or elsewhere. Moreover, the Department will not move any plutonium unless and until all applicable requirements are met.

Thank you for allowing me the opportunity to testify before your subcommittee, and this completes my formal statement.

[The prepared statement of Charles E. Anderson follows:]

PREPARED STATEMENT OF CHARLES E. ANDERSON, PRINCIPAL DEPUTY ASSISSTANT SECRETARY OF ENVIRONMENTAL MANAGEMENT, DEPARTMENT OF ENERGY

Good morning. My name is Charlie Anderson and I am the Department of Energy's Principal Deputy Assistant Secretary for the Office of Environmental Management. I have been involved with Plutonium disposition for a number of years, in this and previous positions. I am also a member of the Department's Nuclear Materials Disposition and Consolidation Coordination Committee (NMDCCC). The Office of Environmental Management is responsible for the safe storage and security of the majority of DOE's surplus non-pit plutonium pending disposition of that material.

The GAO's July 2005 report, *SECURING U.S. NUCLEAR MATERIALS: DOE Needs to Take Action to Safely Consolidate Plutonium,* contains two recommendations for executive action, with which the Department concurs. The first recommendation calls for a comprehensive strategy to be developed for the consolidation, storage, and disposition of DOE's excess plutonium. The second recommendation suggests DOE's cleanup plans be reviewed to ensure they are consistent with the comprehensive strategy for consolidation, storage and disposition.

While the Department believes that consolidation can result in significant benefits with respect to safety and cost savings, any future decisions to do so will be based on the outcome of ongoing evaluations and considerations that will provide the foundation for development of the Nuclear Materials Disposition and Consolidation Coordination Committee's Strategic Plan. The Department remains committed to conduct the disposition of plutonium and any further consolidation of the material in a manner that is good for the environment, safe for the worker, respectful of the taxpayer, and consistent with all applicable statutory requirements. Until the Nuclear Materials Disposition and Consolidation Coordination Committee's Strategic Plan is completed the Department will not make a decision on Plutonium consolidation.

With respect to the first GAO recommendation that the Department develop a comprehensive strategy for the consolidation, storage, and disposition of the Department's excess plutonium, former Secretary Abraham established the Nuclear Materials Disposition and Consolidation Coordination Committee. Secretary Bodman subsequently approved the charter for this Committee. A key responsibility of this Committee is to develop and ensure implementation of a Strategic Plan for disposition and consolidation of special nuclear material. This Strategic Plan will encompass the comprehensive strategy recommended by the GAO.

The principal mission of this Committee is to provide the Department with recommendations on cross-cutting nuclear materials disposition and consolidation planning with the objectives of providing the necessary security for DOE's nuclear materials, identifying paths for disposition, as appropriate, and reducing out-year security and program costs. The scope of material within the Committee's charter includes all of the surplus plutonium owned by the Office of Environmental Management and all surplus non-pit plutonium owned by the National Nuclear Security Administration.

Deputy Secretary Sell recently approved the Mission Need for a plutonium disposition project at the Savannah River Site for plutonium that does not have an identified disposition path; that is, plutonium not suitable for disposition using the currently-designed Mixed Oxide, or MOX, Fuel Fabrication Facility planned to be constructed at the Savannah River Site. The Department's Fiscal Year 2006 congressional budget request includes $10 million for conceptual design of the Savannah River Site plutonium disposition project. As part of this conceptual design, the Department will be evaluating a number of alternatives to meet the disposition objective.

In response to the GAO's second recommendation, I can assure this Congressional Committee that following the completion of the NMDCCC's Strategic Plan, the Office of Environmental Management and its staff will revise and review the appropriate clean up plans to make certain they are consistent with the Strategic Plan and its associated implementation schedule.

I would like to say a few words about the findings contained in the GAO's report, which are based on the premise that Plutonium will be consolidated at the Savannah River Site and have in no way been approved or endorsed by the Department. First, prior to shipping any additional weapons-usable plutonium to the Savannah River Site, the Department will comply with all applicable statutory requirements, including those established by the National Defense Authorization Act for Fiscal Year 2002 and by the National Defense Authorization Act for Fiscal Year 2003. With respect to being able to receive all of Hanford's plutonium, the Savannah River Site's K-Area currently has adequate storage capacity for all of Hanford's plutonium, including the unirradiated fuel rods now stored at Hanford. While additional activities need to be completed, including development of revised safety documentation, shipping package certification and appropriate National Environmental Policy Act, or NEPA, analyses, the Hanford fuel rods can be shipped to and stored intact at the Savannah River Site.

While recent changes to DOE's security requirements make it highly desirable to have as few nuclear material storage locations as possible, the elimination of one facility planned for use of plutonium storage at the Savannah River Site, the Metallurgical Building (Building 235-F), does not complicate our potential storage plans, since K-Area alone now has adequate storage capacity. Furthermore, the Department agrees with the Defense Nuclear Facilities Safety Board in that there are potential safety issues associated with continued use of the Metallurgical Building. Finally, with respect to the GAO finding that "DOE lacks the capability to fully monitor the condition of the plutonium necessary to ensure continued safe storage," the existing plutonium surveillance and monitoring capabilities in the Metallurgical Building are all that are required until 2007 to ensure continued safe storage. Beginning in 2007, the Savannah River Site will have the capability in K-Area to perform all required surveillance and monitoring examinations to ensure safe storage of plutonium at the site.

In closing, it is very important to keep in mind that, while the Department is evaluating options for the safe and secure storage of weapons-useable plutonium, the Department currently has no plans, nor have we made any decisions, to further consolidate such plutonium to the Savannah River Site or elsewhere. Moreover, the Department will not move any plutonium unless and until all applicable requirements are met.

Thank you for allowing me the opportunity to testify before your Subcommittee, and this completes my formal statement. At this time I would be pleased to answer any questions you may have.

Mr. WHITFIELD. Thank you, Mr. Anderson.

Dr. Eggenberger, the Chairman of the Defense Nuclear Facility Safety Board, you are recognized for 5 minutes.

STATEMENT OF A.J. EGGENBERGER

Mr. EGGENBERGER. Thank you, Mr. Chairman. It is indeed a pleasure to present this testimony on the GAO report. With me, I have two of my other board members which I would like to recognize. I have Dr. Jack Mansfield, and Dr. Bruce Matthews here with me also, and this testimony represents the testimony of the Board. And I would request that my testimony be entered into the record, and I will give a summary, a very short summary.

I would like to first say that the Board does agree completely with GAO's assessment, and I would like to bring a little different perspective to the issue. First, in that perspective one needs to know and remember that the Board only gives advice to the Secretary. We are nonregulatory, and we do that through basically two things. One are called recommendations, and the other are called letters.

We put a recommendation to the Secretary of Energy in 1994 which was called Improved Schedule for Remediation in the Defense Nuclear Facilities Complex. Now, the Secretary accepted that recommendation and gave us an implementation plan to implement the recommendation. Included in the implementation plan was a commitment to construct a facility called the Actinide Storage and Packaging Facility.

What that would do was that would allow DOE to stabilize and store in robust containers all legacy materials in the DOE complex. As has been mentioned by the other witnesses and as time went on, things changed within DOE and the plans for disposition have changed.

Another change that occurred was the Actinide Storage and Packaging Facility was also canceled. The thing that was attractive to us, and we believed attractive to the government, was this facility allowed DOE to mass together the legacy plutonium in a safe condition, and in robust containers that would last for a minimum of 50 years.

This would then allow the government to not be required to move quickly without proper background and definition, such that the disposition path could be taken in a manner that was technically very sound. So what we have now, of course, is we are unable to consolidate the material from the various sites to Savannah River, which I believe that in that time period we could have, and we probably moved too fast in that time period in determining exactly what the disposition paths were believed to be.

So as far as DOE being able to come up with a new plan for disposition, I would encourage them to be very careful, to understand the risks, not only technically but politically, and that they look at all possible things that could be done.

People talk about disposition. Maybe it's time, I'm not necessarily saying we should do this, but at least it's time to think about maybe not moving forward with the disposition, and putting the material in a safe and stable condition such that proper decisions could be made. I think we are back at the beginning again, as we were in 1994. This includes, of course, things as chemical processing, and all of other attendant type processes that would enable us to do that.

One last thing, the Board has met with the new DOE committee, the NMDCCC, that will be putting together the recommendations to the Secretary and the Department for a path forward, and our belief is that they are in the beginning stages of their deliberations.

Thank you, sir.

[The prepared statement of A.J. Eggenberger follows:]

PREPARED STATEMENT OF A.J. EGGENBERGER, CHAIRMAN, THE DEFENSE NUCLEAR FACILITIES SAFETY BOARD

INTRODUCTION

Mr. Chairman and Members of the Subcommittee, I appreciate the opportunity to present testimony on the Defense Nuclear Facilities Safety Board's (Board) review of the Department of Energy's (DOE) efforts to consolidate surplus plutonium inventories.

Today's hearing addresses the Government Accountability Office (GAO) audit report, "Securing U.S. Nuclear Materials: DOE Needs to Take Action To Safely Consolidate Plutonium." As indicated in the report, the Board provided substantial technical input to the GAO auditors. GAO found that DOE needed to develop a comprehensive strategy to consolidate, store, and eventually dispose of its plutonium and needed to ensure that its cleanup plans are consistent with its plutonium consolidation plans. The Board agrees with GAO's findings and conclusions that are relevant to the Board's nuclear health and safety jurisdiction.

I would like to summarize the statutory nuclear safety oversight mission of the Board, and then briefly review the Board's recent activities that are relevant to consolidated plutonium storage and disposition. I will also review the Board's Congressionally mandated study of plutonium storage at the Department of Energy's Savannah River Site (SRS) and our suggestions for the safe storage and disposition of excess plutonium.

THE BOARD'S STATUTORY OVERSIGHT MISSION

Congress created the Board as an independent technical agency within the Executive Branch, external to DOE, to identify the nature and consequences of potential nuclear threats to public health and safety at the Department of Energy's defense nuclear facilities, to elevate such issues to the highest levels of authority, and to inform the public. Broadly speaking, the Board provides nuclear safety oversight of DOE's defense nuclear facilities from design through construction, operation (including storage), and decommissioning. The Board is not a regulatory, but an advisory agency with approximately 60 technical staff.

The Board's approach to conducting its nuclear safety oversight mission is to identify to DOE conditions or deficiencies which could adversely affect the public, including workers', health and safety. The Board provides advice and recommendations to DOE primarily by way of letters, reporting requirements, and formal recommendations to the Secretary of Energy. DOE can accept or reject the Board's advice and recommendations. Although DOE's contractors implement most of the nuclear health and safety improvements identified by the Board, the Board works primarily through DOE—both headquarters and site office staff. To date, all Board recommendations have been accepted by the Secretaries of Energy.

The Board conducts its nuclear safety oversight of DOENational Nuclear Security Administration activities at the Los Alamos, Lawrence Livermore, and Sandia National Laboratories; the Pantex Plant, the Y-12 National Security Complex, the Savannah River Site, and the Nevada Test Site. The Board also conducts nuclear safety oversight of DOE's Environmental Management activities at these sites as well as the Hanford Site, Idaho National Laboratory and Idaho Cleanup Project, Oak Ridge National Laboratory, Waste Isolation Pilot Plant, and the Fernald and Mound Sites in Ohio. Operations at DOE's defense nuclear facilities include assembly, disassembly, and dismantlement of nuclear weapons; and maintenance and surveillance of the aging nuclear weapons stockpile. Operations at defense nuclear facilities also include the stabilization and storage of nuclear materials, the deactivation and decommissioning of facilities, and the processing and storage of radioactive waste.

The Board's jurisdiction covers only nuclear safety oversight of DOE's defense nuclear facilities and activities; including the safe storage of plutonium in defense nuclear facilities. As such, some of the issues that are discussed in this hearing, like those directly related to safeguards and security, are beyond the Board's jurisdiction. There may, however, be causal elements associated with these issues that affect nuclear safety and are of interest to the Board. Moreover, there are often important relationships between nuclear safety and security, and between nuclear and industrial safety. Consolidation of nuclear materials is a prime example. It can have both nuclear safety and security components; however, the Board's jurisdiction is limited to nuclear health and safety issues.

BACKGROUND

In the mid1990s, DOE developed a plan for storage of its excess plutonium materials. The inventories of material at the Rocky Flats Environmental Technology Site (Rocky Flats) and SRS were to be stored in a state-of-the-art facility—the Actinide Packaging and Storage Facility (APSF) at SRS. This facility was designed to allow for expansion to accommodate additional nuclear materials from other DOE sites. Advanced monitoring and handling features of this facility would have minimized manual inspection and movement of containers, thereby reducing worker radiation doses and criticality risks.

Additionally, in our Recommendations 94-1, *Improved Schedule for Remediation in the Defense Nuclear Facilities Complex,* and 2000-1, *Prioritization for Stabilizing Nuclear Materials,* the Board encouraged DOE to stabilize and package its excess plutonium into robust storage containers. This action provided DOE time to decide the best course of action for future storage and ultimate disposition of plutonium.

The K-Area reactor facility was built at SRS in the 1950s. The reactor was shut down in the early 1990's. In 1998, DOE decided to modify the facility to accommodate early deinventory of Rocky Flats. This K-Area facility, also known as KAMS (K-Area Material Storage), was intended to be used for a limited time, less than 10 years, until APSF was to become operational.

In 2000, DOE completed a study of plutonium stabilization and storage options. This study assumed that a proposed plutonium immobilization facility would provide a nearterm disposition pathway for DOE's excess plutonium metal and oxides not slated for use in mixedoxide (MOX) fuel. Given the assumed short storage period, the DOE study team concluded it would be more costeffective and timely to modify existing facilities to provide the capability for stabilization and storage than to construct a new facility. Accordingly, the recommendation of the study was to cancel the APSF project and modify Building 235-F (235-F)—originally built in the 1950s—to install a stabilization and packaging capability.

Even though APSF had been designed and excavation begun, DOE canceled construction of the facility in 2001. DOE's decision was based primarily on budget constraints and expectations that a disposition path for the plutonium (MOX and immobilization facilities) would be available in the relatively near future. The immobilization facility was delayed shortly after this decision, and then canceled in 2002. In conjunction with this cancellation, DOE decided that storage of the Rocky Flats plutonium materials in KAMS could extend beyond the 10 years previously estimated.

Since DOE had planned to utilize APSF to provide a means to stabilize, package, store, and conduct surveillance and monitoring of SRS's inventory of plutonium, the decision to cancel APSF left DOE without clear provisions for the safe stabilization and storage of excess plutonium at SRS. To achieve timely stabilization for plutonium at the SRS site, the Board suggested that these materials could be stabilized and packaged efficiently with some minor modifications to the FB-Line. DOE agreed and has now completed stabilization and packaging of the SRS excess plutonium. DOE concluded that storage of the SRS materials could be provided by modifying storage vaults in 235-F and increasing storage capacity in KAMS. In 2002, Congress directed the Board to study the adequacy of plutonium storage at SRS.

CONGRESSIONALLY MANDATED SRS PLUTONIUM STORAGE STUDY BY THE BOARD

In section 3183 of the FY 2003 National Defense Authorization Act, Congress directed the Defense Nuclear Facilities Safety Board to conduct a study of the "adequacy of the K-Area Materials Storage facility (KAMS), and related support facilities such as Building 235-F, at the Savannah River Site, Aiken, South Carolina, for the storage of defense plutonium and defense plutonium materials..." The statute required the Board to:

(1) address—
  (A) the suitability of KAMS and related support facilities for monitoring and observing any defense plutonium or defense plutonium materials stored in KAMS;
  (B) the adequacy of the provisions made by the Department for remote monitoring of such defense plutonium and defense plutonium materials by way of sensors and for handling of retrieval of such defense plutonium and defense plutonium materials; and
  (C) the adequacy of KAMS should such defense plutonium and defense plutonium materials continue to be stored at KAMS after 2019; and
(2) include such proposals as the Defense Nuclear Facilities Safety Board considers appropriate to enhance the nuclear safety, reliability, and functionality of KAMS.

Congress also required both the Secretary of Energy and the Board to submit annual reports on the actions taken by DOE in response to the Board's proposals. The first annual report was required to be submitted six months after the Board's study was submitted. Subsequently, the Board has submitted a 2004 and 2005 annual report to Congress pursuant to this statute.

BOARD PLUTONIUM STUDY FINDINGS

In our study, *Plutonium Storage at the Department of Energy's Savannah River Site,* dated December 1, 2003, the Board made proposals concerning DOE's plutonium disposition program, the suitability of 50-year-old facilities planned for storing plutonium at the SRS, and the remote monitoring and retrieval of plutonium. The Board proposed safety upgrades to ensure the nuclear safety, reliability, and functionality of the existing facilities (KAMS and 235-F) proposed for plutonium storage. The Board also proposed that DOE expedite the development of a complete, wellconsidered plan for the final disposition of all excess plutonium to minimize unnecessary extended storage of plutonium at SRS. Even with a sound disposition plan, excess plutonium is expected to be stored for several decades at SRS; therefore, the Board additionally proposed that DOE conduct a new study of available options for the storage of plutonium at SRS.

In April 2005, DOE decided to consolidate the excess plutonium currently at SRS into the KAMS facility and not utilize 235-F for extended storage. This decision obviates the need for nuclear safety upgrades to 235-F related to extended storage.

The Board considers the KAMS facility to be a robust structure that can be made suitable for storage by establishing an appropriate fire protection system and eliminating unnecessary combustibles. DOE has agreed to remove unnecessary combustibles and has recently directed that needed upgrades to the facility's fire protection system be made. The combination of these actions and the robust packaging containers required for storage in KAMS, provides a suitable facility for storage of plutonium. To meet existing DOE requirements for extended storage, DOE will need to add the capability to monitor, stabilize and repackage plutonium in this facility. DOE plans for this activity are in progress.

CURRENT STATUS OF PLUTONIUM STORAGE AND CONSOLIDATION

In the Board's 2004 and 2005 annual follow-up reports to Congress on the plutonium storage study, the Board stated that DOE had not established a consistent, well-considered plan for storage and disposition of excess plutonium. Rather, DOE's storage plans continue to change. DOE has been unsuccessful in consolidating excess plutonium at SRS. DOE has directed that the Hanford Site plan to store its excess plutonium on site through 2035. DOE's laboratories must also continue to store excess plutonium. Contributing to consolidation difficulties are inconsistencies between Hanford and SRS as to how the plutonium must be packaged before shipping to SRS (i.e., unirradiated Fast Flux Test Facility fuel at Hanford Site). Specific actions to accommodate this new direction for extended storage of excess plutonium at various sites and to address packaging have not been formalized by DOE and have not been evaluated by the Board. However, this strategy raises potential questions about the nuclear safety of options being considered by DOE to store plutonium in areas never intended for such storage.

For extended storage, consolidation of excess plutonium into a single, robust facility specifically designed for storage is logical from a nuclear safety perspective. Accordingly, the Board has advised DOE to consider broader alternatives for safe and secure storage of its excess plutonium. If unable to consolidate plutonium at existing SRS facilities, DOE should consider other locations for consolidation of plutonium. Options include consolidation in a new facility, specifically designed for such storage, or consolidation in an existing facility that has been determined suitable for extended storage.

DOE's current disposition strategy for excess plutonium consists primarily of processing into mixed-oxide fuel or vitrifying into lanthanide borosilicate glass for disposal. A small quantity of excess plutonium is to be disposed of as waste either at the Waste Isolation Pilot Plant or through the SRS high level waste system. As envisioned, the vitrification process would be established in areas of the K-Reactor facility at SRS. This vitrification process is preliminary and still years away from being realized.

Given DOE's decision to ultimately dispose of its excess plutonium, the Board advised DOE to consider additional alternatives for its disposition strategy, including the potential for incorporating more of the material into MOX fuel. In lieu of pursuing the vitrification project only, DOE has recently approved the mission need for a plutonium disposition project. This project includes developing disposition alternatives that take into account other ongoing or planned plutonium processing activities. This appears to be an appropriate reconsideration of the path forward on plutonium disposition.

The two Board proposals from its recent 2005 follow-up report to Congress, namely that DOE consider broader alternatives for storage and that DOE consider additional alternatives to disposition, are consistent with the GAO report findings.

In early 2005, DOE formed a new broadly chartered group—the Nuclear Materials Disposition and Consolidation Coordination Committee—comprising senior DOE management personnel, which may provide the strategic planning needed. This group is to provide a forum to perform crosscutting nuclear material disposition and consolidation planning for DOE. This is a positive development but the committee does not have a clearly identified set of goals, objectives or schedule nor has this committee, to date, provided any real strategic planning that is obvious to the Board. DOE continues to develop new plans and alternative plans since 1995 but has not implemented any of them to date.

Mr. WHITFIELD. Dr. Eggenberger, thank you. And since your first discussions of this issue in 1994 with the Secretary of Energy at that time, how much progress do you feel has been made by the DOE?

Mr. EGGENBERGER. I would have to say, sir, that a tremendous amount of progress has been made for the following reason. The first reason is things were in disarray at that point in time. The material has been amassed. We know where the material is. Much of it has been stabilized, and the container to put the material in has been designed, has been fabricated and is in use at the Savannah River site and other sites for storage at this point in time. It is robust.

Again, 1994 is 10 years or so ago, and DOE has a program for monitoring the can, and doing surveillance on the can to see if it will take its 50-year life that we—that the Department has designed it for.

So in that sense we believe that they have been making a lot of progress. The idea of amassing the material all in one place and disposing of it, if you wish, or holding onto it until you are ready to come up with a plan that provides all of the necessary elements for disposal, we haven't done—they have not done as good on that, sir.

Mr. WHITFIELD. Mr. Aloise, from your information and studies that you have conducted on this whole issue, would you go over again with us from your perspective what the major problems are that still face DOE in making a final decision about this?

Mr. ALOISE. Well, in our view, one of the major problems is the lack of this complex-wide coordinated plan.

Mr. WHITFIELD. The lack of what?

Mr. ALOISE. A complex-wide coordinated plan, which looks at all of the individual sites and the problems and issues at each site, and in one plan kind of formulizes a decision of what we are going to do and what is the path forward.

And I think without that, we are not going to—even though I agreed they have made a lot of progress to date, without that final plan, little progress from here probably will be made.

Mr. WHITFIELD. So from your perspective and from what you know, it does not appear that they are in a position to conclude this plan in the immediate future?

Mr. ALOISE. Not in the immediate future. Not that we see, no.

Mr. WHITFIELD. What do you mean by immediate?

Mr. ALOISE. Well, I think the committee is just starting their work on this looking at that, and we really do not have a schedule or timeframe or have any idea when this plan is going to happen.

Mr. WHITFIELD. Now, Mr. Anderson, you mentioned in the Defense Authorization Acts of 2002 and 2003, I believe, and that they required certain things from DOE before—were those requirements applicable only to the Savannah River site or to any site, the prohibitions or requirements in those authorizations acts of 2002, 2003?

Mr. ANDERSON. They are applicable in the context of the Savannah River site, in the context of receiving plutonium or maintaining storage of weapons usable plutonium in the State of South Carolina.

Mr. WHITFIELD. Now, what is specifically required under those acts?

Mr. ANDERSON. It required that certain performance objectives be met for the mixed oxide fuel fabrication facility. And if not, then there is a planning for—suspension of any further shipments of plutonium into the State and planning for the potential removal of plutonium from the State of South Carolina.

Mr. WHITFIELD. Now, and——

Mr. ANDERSON. It also requires a corrective action plan if those performance objectives for the mixed oxide fuel fabrication facility are determined they can't be met, and that is consistent with the letter by the Secretary dated August 15th. We are initiating all three avenues of that, the corrective action plan for mixed oxide fuel fabrication facility, and also an evaluation of plans for the removal of any plutonium if those performance objectives can't be met, and suspension of the plutonium shipments into the State.

Mr. WHITFIELD. Now, you made it quite clear that DOE has made no decision about a location for consolidation of any of this material. That's correct?

Mr. ANDERSON. Yes, sir.

Mr. WHITFIELD. So this Defense Authorization Act applies only to the Savannah River site. So if you are considering other sites, it would not be applicable to any of those sites. Is that correct?

Mr. ANDERSON. That's correct.

Mr. WHITFIELD. Okay. And the Nuclear Materials Disposition and Consolidation Coordination Committee, how many people are on that committee?

Mr. ANDERSON. There is about 8, 8 to 10, and of course supported as we deem necessary with additional subject matter experts.

Mr. WHITFIELD. Who is the Chairman of that committee?

Mr. ANDERSON. It was the Senior Policy Adviser for National Security to the Secretary, but she has taken a different position. So right now it is being headed up by the Under Secretary for ESE and Ambassador Brooks.

Mr. WHITFIELD. Now, who was the Chairman? I understand she resigned recently?

Mr. ANDERSON. Yes, sir. That is correct. Senior Policy Adviser to the Secretary was the Chairman. The steering committee, the senior executive steering committee was made up of her and Ambassador Brooks and Under Secretary Garman.

So at this point in time, as far as chairing the committee, it is going to be co-chaired with Under Secretary Garman and Ambassador Brooks.

Mr. WHITFIELD. So you have co-chairs now?

Mr. ANDERSON. Correct.

Mr. WHITFIELD. How long has this been in effect?

Mr. ANDERSON. About a week.

Mr. WHITFIELD. One week. How often does the committee meet?

Mr. ANDERSON. I would give it on the order of about once a month or every 6 weeks. But it is typically as necessary to either gather or analyze data, and a large portion of the deliberations so far have been making sure we had all of the data identified and, as Dr. Eggenberger indicated here, to make sure that we identified the risk and the potential options.

Previous efforts along these lines have resulted in projects that we ultimately then later decided not to do, because we believed that all of the risk and any options have not been completely evaluated, and the Secretary has made it very, very clear that he wants us to do a complete evaluation so that we know what the risks are so that we can make commitments that we will fully execute.

Mr. WHITFIELD. Now, you serve on the committee?

Mr. ANDERSON. Yes, sir.

Mr. WHITFIELD. You represent what, the Office of Environmental Management?

Mr. ANDERSON. I represent the Office of Environmental Management. I have previously also worked with the NNSA. So I bring some relationship with this from both areas.

Mr. WHITFIELD. Now the charter requires that there be what, weekly telephone conference calls, is that correct, or monthly teleconference calls?

Mr. ANDERSON. I don't recall that in the charter.

Mr. WHITFIELD. From your perspective, you have been a member of this subcommittee since its formation?

Mr. ANDERSON. Since soon after its formation, yes, sir.

Mr. WHITFIELD. So sort of in layman's terms if you were speaking to the Rotary Club down in some little city down in Kentucky and you were explaining to them what you were doing, what would you say to them as far as your progress and when you expect to be able to reach a final decision?

Mr. ANDERSON. As far as the progress we have made, we have identified all of the materials. We think we have identified all of the options for consolidation, and I will say many of the barriers, the roadblocks to evaluating those, the risks that we would have.

We are trying to beat the bushes, if you will, to make sure that we have identified all of those risks. We do not want to repeat some of our problems in the past. One of the aspects of this is while we have identified—obviously we know where our material is and we have identified that material, is making sure we fully understand the condition of the material.

A large portion of that material has been stabilized into these robust containers. We refer to them as 30-13 containers, but there is material such as the fuel at the Hanford site which has not yet been stabilized in that material, in that form. So we have got to make sure that we understand completely what those characterizations are and then what the options are for consolidating the material, whether we would put it in those robust containers or store it in some other form.

Mr. WHITFIELD. And if you were looking at a date in the future that you would hope to be able to reach some conclusion about this, when would that be?

Mr. ANDERSON. I think it is going to still be some time because of the cost numbers that—we also have to make sure. I mean I— with some of our cost estimates in the past on the project, the Secretary has been very clear to us that he wants to make sure that we have very firm estimates for our options. So when we are talking about evaluating our options and we review it with him, that we have a good basis for that. Obviously some of the options have very detailed estimates for them, and some of them don't. So we are having to go back and look and make sure that we understand what the basis is for those numbers.

A lot of the cost numbers we look at are based on the Design Basis Threat, which obviously was upgraded a little over a year ago. And that is—that has thrown a new wrinkle as far as, you know, what our cost numbers are.

Mr. WHITFIELD. Mr. Aloise I think referred to some $85 million figure, or was that Dr. Eggenberger? What was that $85 million figure?

Mr. ALOISE. The $85 million is the additional storage cost for the material at Hanford because they are not able to consolidate it, per year.

Mr. WHITFIELD. That is per year. And the costs of consolidating, we really do not have any idea what that figure might be. Is that correct, Mr. Anderson?

Mr. ANDERSON. The actual cost of consolidating we don't yet know, because we have to analyze whether we already meet the Design Basis Threat, whether that cost is clear or whether we have to do additional work in Design Basis Threat to do that. For instance, in many of the sites we would have to implement Design Basis Threat requirements that we weren't planning to before.

Mr. WHITFIELD. Thank you. At this time I recognize Mrs. Blackburn for 10 minutes.

Mrs. BLACKBURN. Thank you, Mr. Chairman. Thank you all for being here with us this morning. And Mr. Anderson, I think I would like to stay with you. I find it interesting that you are so uncertain about the amount of time it's going to take and the cost you will bear and the risks that are before you when you talk about doing a complete evaluation.

And speaking to the chairman's point, I think what we want to hear from you all is an orderly process, how you plan to proceed, and what your timeline is going to be, your best estimate.

And if we have that type information, I think that would be helpful. If as you work through, if you all could provide that to us in writing so that we will have it for follow-up, if you don't mind, sir.

Okay, let me ask you something else. Speaking of the risks, I know that other foreign nations, primarily France and Germany and one other, recycle their plutonium. And in, I think it was 2002, the administration, you all had planned on a recycle project. Am I correct on that?

Mr. ANDERSON. We are planning on a disposition project, the mixed oxide fuel fabrication facility, and an ultimate irradiation of that material in commercial reactors.

Mrs. BLACKBURN. Okay. And why have you not moved forward with that recycling project?

Mr. ANDERSON. We have continued to move forward. There was some design and licensing issues that they have been working through on that project.

Mrs. BLACKBURN. Okay. And could you please tell me what countries you are looking at for lessons learned as you look at moving forward on that project?

Mr. ANDERSON. I would like to get back with you. I know it includes the ones that you have mentioned there, but to provide you a complete list, we would like to get back with you on that.

[The following was received for the record:]

While the Department does not currently "recycle" plutonium, we are moving forward with the MOX program. This program has some of the aspects found in similar programs in France, Germany, Belgium, and Russia.

The U.S. MOX facility that will be built at the Savannah River Site is based on MOX fuel fabrication technology that is being used at Cogema facilities in France. As a result, we are able to take maximum advantage of Cogema's operating experience for the U.S. facility. MOX fuel fabrication facilities have operated successfully in Europe for over four decades, and MOX fuel is being used successfully in nuclear power facilities around the world. This program will put surplus weapons-useable plutonium into mixed plutonium-uranium oxide fuel which will be irradiated in certain specified commercial power reactors. The MOX spent fuel will be as resistant to nuclear proliferation as is commercial spent fuel and cannot readily be used in nuclear weapons.

Mrs. BLACKBURN. Okay. And then also another thing that I would be interested in knowing would be how much you think it would save us, both as a cost-benefit analysis, as an environmental analysis, what you think the savings on each of those fronts would be. And, if you have—if you have looked at that, and if you have done that type work. Is that work in the process?

Mr. ANDERSON. That is the work that is in the process. As I mentioned earlier, there is some uncertainty around some of these numbers and the different options because of the level of understanding about the options that we have there.

We understand it would be a significant amount of savings in consolidating our material. But that is one of the reasons we want to pursue that. But we haven't really finalized a number that we could commit to at this point.

Mrs. BLACKBURN. Okay. And then you mentioned a new facility, and building a new facility. And if Congress were to give the money and the authorization for a new facility, then how long would it be, in your best estimate? How long would it be to complete that facility, and then to transport the plutonium to that site to consolidate it?

Mr. ANDERSON. Again, without really going through that, it is hard to give you that. If I give you the kind of number like that, we would be back to where some of our failures have been in the past. And that is where, you know, when you look back to the packing and storage facility and some of our previous projects, you know, we have laid out a timeline for those projects, then as they start to stretch out it affects both the validity of that and our ultimate goal of consolidating that material.

Mrs. BLACKBURN. Okay. And at present the total number of facilities that you're planning to use to store plutonium are what? What was that total number? I think you hit that earlier?

Mr. ANDERSON. Currently in the environmental management program we have consolidated our plutonium down to two facilities, one at the Savannah River site and one at Hanford. The NNSA also has material at Los Alamos, Lawrence Livermore and Pantex.

Mrs. BLACKBURN. Okay. Mr. Aloise, the plutonium at the Hanford site, can it be changed into a form that can be transported to the Savannah River site?

Mr. ALOISE. Well, it is our understanding that it could be shipped in its present form. But there is no approved Department of Transportation container to do that yet. But it can be—it could be stored in its present form as well as according to DOE officials. If they were to package it like the rest of the plutonium, we are told that it would be an estimate of another 1,000 containers to hold that material.

Mrs. BLACKBURN. Okay. And how is Yucca Mountain a part of that process?

Mr. ALOISE. Well, Yucca Mountain is being talked about as the final disposition path. But, as we all know, that has not been decided yet.

Mrs. BLACKBURN. And you see it as less risk to truck this across the country than to develop a recycling program?

Mr. ALOISE. No, ma'am, I don't. That is something we are calling for in DOE's plan to determine what that risk is.

Mrs. BLACKBURN. All right. Thank you, Mr. Chairman. I yield back.

Mr. WHITFIELD. Thank you, Mrs. Blackburn. At this point I recognize Dr. Burgess for his 10 minutes of questioning.

Mr. BURGESS. Well, thank you, Chairman. I thank the witnesses for being with us this morning. The ability to transport that material across country, of course, is no small feat. And having seen, having had an opportunity to visit some of your facilities out in Albuquerque that does some of the transport, I was very impressed with the care of the people who worked there do take with that transport.

But still it does strike me that going all of the way out to Savannah, Georgia, and then coming all of the way back to Yucca Mountain at some point in the future doesn't seem like a best use of manpower and resources.

I get the impression from the GAO report, Mr. Aloise, that you're not satisfied that we are making satisfactory progress toward that end. Is that a fair statement?

Mr. ALOISE. That's a fair statement. We would have liked to have—when we began this review we expected to see a plan. But what we saw instead were individual site plans which in fact conflicted with each other. So not much progress is going to be made, we don't believe, until a plan is developed.

Mr. BURGESS. And is this a problem with management or a problem with oversight? Is it a problem where you need additional legislation, relief from legislation? What can this committee do to help that?

Mr. ALOISE. I think that question can be better directed to Mr. Anderson. But I believe they have begun the process with this committee, I mean, this planning committee.

Mr. BURGESS. I would like to give Mr. Anderson a chance to respond to that. I have never been a big believer in committees, but I sit on a committee now. So help us with that.

Mr. ANDERSON. Certainly. I think with Secretary Bodman coming on board and making sure that we are looking at this from an integrated fashion, both across the NNSA side and the rest of the Department, is one of the major steps that we see in coming up with a comprehensive strategy. Prior to that there were a lot of stovepipes, a lot of barriers between the two organizations.

You know, with that in mind, there—in any of the paths there may be some legislative relief or something, but I mean we are trying to identify those as risks.

In other words, looking at any barriers, if we look at a technical basis, we are trying to say, if we were going to consolidate at any particular site, is there some legislative support we need or is there some legislative barrier, legal barrier that we need to have some relief on.

So it really depends then on which one of those answers come through.

Mr. BURGESS. Well, I'm a big believer in eliminating stovepipes. Do you feel that the organization of this committee, the MNDCCC committee, that almost sounds like a stovepipe, doesn't it? Do you think that's going to help?

Mr. ANDERSON. Absolutely. I mean, it is the form, it is designed to set up the form for all of the organizations that have all special nuclear materials to come together and discuss that disposition and consolidation, not just the plutonium disposition, plutonium materials, but any of our special nuclear materials since they were required for security.

Mr. BURGESS. Well, it won't help if we do not do the monthly teleconference calls and the things that were outlined in the charter.

Mr. Chairman, can I ask that we be kept apprised as to the performance of those? I mean, those are performance measures that we can monitor from this committee. Would that be out of line to ask for that?

Mr. WHITFIELD. No, that is certainly true, and we would like to maintain contact with you through our committee staff on that and are quite interested in a solution to this, and if we could set up a system where we could have some monthly contacts or at least quarterly with our staff, it would be quite helpful.

Mr. ANDERSON. Absolutely.

Mr. BURGESS. Thank you, Mr. Chairman.

Well, Mr. Anderson, for each year—my understanding is for each year that the Department of Energy is unable to remove plutonium out of the Hanford site to the Savannah River site these ongoing storage costs will be at least $85 million annually, give or take. Is that correct?

Mr. ANDERSON. That is correct.

Mr. BURGESS. The continued storage will delay the Department's goal of accelerating cleanup at Hanford. Now a February 15, 2005, Department of Energy memo that I have here states that the plutonium at Hanford may remain there until the year 2035.

Do you think that in a time of constrained budgets that the Department of Energy should be doing everything it can to remove the plutonium at Hanford as quickly as possible?

Mr. ANDERSON. That is correct. The particular memo that you are referring to was to be used as just a planning basis for that site, because they had run up against their earlier plan, and they realized they would not have the plutonium off of the Hanford site by end of 2006.

Mr. BURGESS. I'm not absolutely sure the math is correct, so someone please check me. But if the plutonium remains at Hanford for 30 years that will cost the taxpayers $2.5 billion. Is it reasonable that we should keep the plutonium at Hanford for so long?

Mr. ANDERSON. We do not believe so. That is one of the reasons we are looking hard at a comprehensive strategy for consolidation of the material.

Mr. BURGESS. What will be the impact if you can't get it done? If it doesn't happen, what would be the impact from failing to remove the plutonium from Hanford to the State of Washington?

Mr. ANDERSON. Well, our failure to be able to consolidate plutonium, whether it is removed from Hanford or consolidated there or whatever site that we would end up with, would be a tremendous cost to the taxpayers. That's why we definitely want to do that.

Mr. BURGESS. Okay. I thank you for your candor.

Dr. Eggenberger, just in the little time that I have left, we talked about the NMDCCC and in your testimony you state that this committee was created by DOE, is a positive development, but the committee does not have a clearly defined set of goals, objectives or schedule, nor has the committee to date provided a real strategic planning that is obvious to the Board.

Again, I will just underscore that if it is proper, I would hope that you would keep this committee informed as going forward, that these performance measures are in fact met. Because I do not know as I sit here of any other way of telling that we are in fact on track for this.

Is that a reasonable request, sir?

Mr. EGGENBERGER. From us?

Mr. BURGESS. Yes.

Mr. EGGENBERGER. Yes. Absolutely.

Mr. BURGESS. How much time do you think it should take for the committee to complete a strategic plan for all of the nuclear materials consolidation?

Mr. EGGENBERGER. I don't know the answer to that, but I can talk about it a little bit.

Mr. BURGESS. Okay. We do that here in Washington.

Mr. EGGENBERGER. The important thing I think is to define the problem. What they are attempting to do is what we call a large complex system. And it is. It is very complex. There is lots of constraints on it. It is necessary that these be defined well. They have not been in the past.

One seems to get a bit cynical as one gets older, and I don't know why they should be able to do a good job of it now. It's a very difficult thing to do. But I hope they can. The one very important thing is I hope that they look at the paths forward in the light of special interests, political agendas, because these all affect the plan and can have an enormous affect on the output of it.

And so the risks that are posed by these need to be considered very carefully. The technical aspects also in the past have been considered in a rather superficial manner, and I would hope that they are able to look at them in depth.

And given that, I would say, to come up with this, it is at least a year or 2-year activity. That is my view.

Mr. BURGESS. I would say, well, temptation may devolve into cynicism. I hope you will not—this is terribly important work. And what we have seen at Los Alamos with the reduction in the nuclear arsenal—which I believe is appropriate and now we have to do the correct thing with those materials as they are no longer needed for national defense.

Mr. EGGENBERGER. That is true. And we believe, as I said earlier, that there have been a lot of good things done since this thing first started taking shape in 1994. And so I hope that this time it's done, that we don't have to repeat this. We've repeated it now two or three times, and it certainly makes our job easier when the work gets done.

Mr. BURGESS. Yes, sir. So that's 11 years, from 1994 to now. Mr. Anderson, you heard Dr. Eggenberger say 1 to 2 years. Do you think that's a reasonable timeframe?

Mr. ANDERSON. For a complete evaluation, yes, that could be a reasonable timeframe.

Mr. BURGESS. Could we push for one?

Mr. Chairman, my time has expired. I will yield back. Thank you.

Mr. WHITFIELD. Thank you, Dr. Burgess. At this time I recognize the gentleman from Washington, Mr. Inslee, for 10 minutes.

Mr. INSLEE. Thank you.

Mr. Aloise, I represent a district just north of Seattle and I used to represent the district where Hanford is located. So, just by way of introduction.

What is your description of the general reasons why it does not appear that there is, at least from our perspective, adequate progress for completing this plan for permanent disposition? And I missed some of your earlier testimony. My apologies.

Mr. ALOISE. I mentioned earlier, actually when we started this review we were hoping to find a plan to analyze and see what DOE's path forward was. And instead, there were individual site plans which actually conflicted with each other. I can't tell you the exact reason why a plan hasn't been developed yet, but we are hopeful that one will be soon.

Mr. INSLEE. So, Mr. Anderson, do we need a central plan? And if we do, why don't we have one?

Mr. ANDERSON. We absolutely do. We are pursuing that. As I mentioned earlier, one of the things that we would be looking at there is to make sure that this comprehensive plan does address all of the risks and barriers so we do not repeat some of the mistakes that we have in the past where we've started down a particular project, found out we had a barrier that was not able to be overcome, have to stop that project and restart down a different path.

Mr. INSLEE. So the obvious question is if Mr. Anderson says he has a plan, why can't Mr. Aloise find it?

Mr. ANDERSON. We are developing a plan for that. We do not have that plan completed yet.

Mr. INSLEE. I misunderstood. I'm sorry. So what is the date for completion of that plan?

Mr. ANDERSON. We do not have a date to complete the plan at this point. The Secretary has been very clear to tell us he wants to make sure whatever time it takes, that it is complete, that it advises all the risks, and that we can make commitments that the Department will live up to.

Mr. INSLEE. Could you give us some parameters? Let me tell you why I'm asking. There is some cost associated with this. There's some security issues associated with this. There's some long-term frustration in the State of Washington for failure to move forward. I would think you could give us some parameters of when that job could get done.

Mr. ANDERSON. The prior question was looking at a reasonable timeframe, you know, within a year or 2, and obviously the urgency to make that closer to a year; but again, we will pursue this as quickly as we can, but making sure that we do have the right numbers. A lot of those numbers are the cost and the impacts that we will have at the different sites.

Mr. INSLEE. What would have to happen for Congress or somewhere else to make that a year and a half? If we wanted to give you a deadline of a year and a half, what would it take to accomplish that?

Mr. ANDERSON. I'm sorry, I'm not sure I understand the question.

Mr. INSLEE. Would it take an additional appropriation? Would it take a congressional mandate? What would it take to get that job done and make sure it's done in a year and a half?

Mr. ANDERSON. We don't believe it will take anything else from Congress to get that done in that timeframe.

Mr. INSLEE. Mr. Aloise, what do you see as downsides of delay in this process?

Mr. ALOISE. Well, there are obviously the additional storage costs, the security costs, and Hanford's inability to finish its cleanup. They wanted to accelerate the cleanup there, and that won't happen until we can get this material consolidated and off premises.

Mr. INSLEE. Mr. Anderson, as far as developing this plan, do you have any constraints in that regard? Is that one of the reasons for delay in getting this job done?

Mr. ANDERSON. No, absolutely not. I guess the force and function, I would say, that's made us pursue this plan which I believe will be successful than in the past, is Secretary Bodman's emphasis on a complete comprehensive group. The creation and charter for the nuclear materials disposition and consolidation coordinating committee emphasizes a form for all people who have these type of materials to come together to make sure we know what the materials are, we evaluate what the options for disposition are, and we look at how to do the proper storage.

One of the things that the GAO has indicated is that during their review what they saw were individual plans, in and of themselves not necessarily bad, but they weren't integrated, and it wasn't a higher-level comprehensive plan, and it was driving input to those individual site plans. So that's the reason for really pursuing a comprehensive plan.

Mr. INSLEE. Just one comment, you know we built the whole Hanford facility and devised this entire new technology in about 2 years, and now we are having difficulty within that period of time coming up with a plan to consolidate the wastes that were generated from that. We just urge you to expedite this process. Thank you.

Mr. WHITFIELD. Thank you Mr. Inslee.

Mr. Anderson, in April of this year, DOE decided to consolidate all the plutonium currently at the Savannah River site in one facility, Building 105-K. And I know our committee staff were down there visiting that site recently, and they reported that there were significant safety system upgrades needed as well as monitoring the surveillance capabilities and so forth before it could really be prepared. In your statement you note that beginning in 2007, DOE will upgrade the facility to perform all required surveillance and monitoring examinations to ensure safe storage of plutonium at the site.

Do you already have an approved plan and the necessary funding to install those necessary upgrades to Building 105-K?

Mr. ANDERSON. Actually we had a project that was approved to put those upgrades in the metallurgical building, and we're working with the Congress to modify that so that those—we're not putting any of those upgrades in. I refer to it as 235-F metallurgical building, but those upgrades now will go into the 105-K and perform the safety upgrades.

Mr. WHITFIELD. Do you have any idea how much the cost will be to upgrade that building?

Mr. ANDERSON. It's in the project data sheet that's a part of that, but the total upgrades are around $100 million.

Mr. WHITFIELD. Do you have sufficient funding now for that?

Mr. ANDERSON. Currently we have a request in to move some funding to be able to support that. That has not yet been approved.

Mr. WHITFIELD. Okay. Mrs. Blackburn in her questions talked quite a bit about recycling, and it's my understanding that recycling of the so-called pit plutonium or the weapon-grade plutonium is normally what we do on recycling. That's the material that we recycle. Is that a part of your plan that you're looking into is recycling that material?

Mr. ANDERSON. Actually, a key cornerstone of the disposition is the mixed oxide program, which is to take the plutonium and convert it into a mixed oxide fuel and irradiate that in commercial reactors.

Mr. WHITFIELD. Of course you had mentioned Europe, but I am supposing that Russia would be the other country that would have as much plutonium, or maybe more than we do. Would that be accurate?

Mr. ANDERSON. The major driver for that program is a like disposition of their material also along that same path.

Mr. WHITFIELD. Do you all have quite a bit of dialog with them about that subject?

Mr. ANDERSON. Quite a bit.

Mr. WHITFIELD. Okay. How would you describe the progress that they're making on this issue?

Mr. ANDERSON. There's been some issues to work along that, both in that regard and design and some potential funding issues that are there, but it is progressing.

Mr. WHITFIELD. But you do think that from Secretary Bodman's standpoint that this is a priority, to finish this plan, to come up with a solution to this issue? You view it as one of his top priorities?

Mr. ANDERSON. Absolutely.

Mr. WHITFIELD. Okay. Before we conclude the hearing I am going to ask unanimous consent, since there is no one here but me to object—I am not going to object—I am going to submit into the record the charter of the nuclear materials disposition and consolidation and coordinating committee, the requirements of that committee. Also a memo from Paul Golan regarding the implementation to Keith Kline regarding this design basis threat policy; and then also the memo from Clay Sell to James Rispolay—is that the correct pronunciation—regarding the approval of the mission need for a plutonium disposition project.

And the committee does intend to stay in touch with you all on this issue, with Secretary Bodman and you, Mr. Anderson. And we do thank you for testifying today. We appreciate the update, and we will be following with great interest. And if there is anything that we can be helpful with, we want to be; and, of course, we may be coming forth with additional questions.

The record will remain open for 30 days in case any member has any additional material they would like to put in. With that, the hearing is concluded.

[Whereupon, at 10:12 a.m., the committee was adjourned.]

[Additional material submitted for the record follows:]

Response for the Record by Charles E. Anderson, Principal Deputy Assistant Secretary, Office of
Environmental Management, Department of Energy

QUESTION FROM CHAIRMAN WHITFIELD

According to the February 17, 2005, Memorandum creating the Nuclear Materials Disposition
and Consolidation Coordination Committee (NMDCCC), "the Executive Steering Committee
will meet twice annually, or more frequently, if necessary. The NMDCCC will meet more
frequently in quarterly meetings and via monthly teleconference calls to assure effective, timely
exchange of information …"

Q1. Please provide a list of dates for each meeting or teleconference the Executive Steering
Committee or the NMDCCC that has occurred to date. Please also provide a list of dates
for planned future meetings.

A1. Since its creation, the Nuclear Materials Disposition and Consolidation Coordination

Committee (NMDCCC) has thus far met three times: on March 25, May 6, and July 27,

2005. Both the May 6 and July 27 meetings were also Executive Steering Committee

meetings. Additionally, there were meetings on December 3, 2004 and February 16, 2005,

of the Nuclear Materials Consolidation Coordination Committee, essentially the same group

as the NMDCCC. Subsequently, disposition was added to the charter of the committee, as

indicated by the February 17, 2005, memorandum. The next meeting will be held in

December 2005. As the NMDCCC gains working experience, its charter may be revised to

foster better efficiency. The Department will inform Congress if this happens.

QUESTION FROM REPRESENTATIVE BLACKBURN

Q1. From the testimony, it looks like that DOE had a plan in the mid-1990s for storage of excess plutonium materials, and completed a study of storage options in 2000. This study concluded that it was cheaper and more efficient to modify a 1950 facility rather than construct a new facility. How much difference in cost and time was this alternative?

A1. The 2000 study, "Evaluation of Savannah River Plutonium Storage and Stabilization Options," recommended utilizing, with modifications, existing Savannah River Site facilities rather than construction of a new facility. The total cost to modify existing facilities was estimated to be $120 to $280 million, consisting of $20 to $30 million to increase storage capacity in the K-Area facility and $100 to $250 million to install a stabilization and packaging capability in the Metallurgical Building. Implementation of this option would take up to five years. The total cost to complete construction of a new storage facility to hold 5,000 standard plutonium containers, including a stabilization and packaging capability, was about $380 million, while the total cost to complete construction of a new storage facility to hold 10,000 standard plutonium containers, including a packaging and stabilization capability, was estimated to be $490 to $590 million. Both new facilities would be operational approximately five years after authorization.

Q2. What process did DOE use to decide that the K-Area facility could store plutonium materials beyond the 10 years previously estimated?

A2. The Department prepared a revised safety analysis of the facility and the specific plutonium storage configuration in the facility. This analysis was documented for the public in a Supplement Analysis, DOE/EIS-0229-SA-2, dated February 2002. The results of that analysis demonstrated that safe storage of plutonium in K-Area can continue beyond 10 years, pending disposition. This Supplement Analysis is referenced in an Amended Record of Decision, "Surplus Plutonium Disposition Program," that was published in the *Federal Register* on April 19, 2002.

The Supplemental Analysis provided the analysis of plutonium storage beyond 10 years. DOE implemented a K-Area Structural Assessment Program (KSAP) to determine the condition of Building 105-K and the Department plans to conduct Structural Assessment every five years to assure safe storage at KAMS. The Supplemental Analysis concluded that plutonium materials can be safely stored for up to 50 years.

The Deputy Secretary of Energy
Washington, DC 20585

SEP 0 6 2005

MEMORANDUM FOR JAMES A. RISPOLI
        ASSISTANT SECRETARY FOR ENVIRONMENTAL
        MANAGEMENT

FROM: CLAY SELL
              DEPUTY SECRETARY

SUBJECT: Approval of Mission Need (CD-0) for a Plutonium Disposition
              Project.

Based on the recommendation from the Energy Systems Acquisition Advisory Board
conducted on August 5, 2005, I approve the Mission Need (Critical Decision-0) for a
Plutonium Disposition Project for plutonium without an identified disposition path. The
approved cost range for this project is $300M-$500M with a projected completion date of
2012.

In support of Critical Decision -1, EM, in collaboration with NNSA and other
Departmental stakeholders, is to develop disposition alternatives that take into
consideration other ongoing or planned plutonium processing activities. The alternatives
analysis should consider the modification of existing facilities as well as the use, or
modification of facilities under development.

cc: Under Secretary for Energy, Science and Environment
     Under Secretary for Nuclear Security/Administrator for National Nuclear Security
     Administration
     Director, Office of Civilian Radioactive Waste Management
     Director, Office of Nuclear Energy, Science and Technology

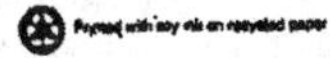

Department of Energy
Washington, DC 20585

February 17, 2005

MEMORANDUM FOR SAMUEL W. BODMAN

THROUGH: LINTON F. BROOKS, UNDER SECRETARY FOR
     NUCLEAR SECURITY

     DAVID K. GARMAN, ASSISTANT SECRETARY
     FOR ENERGY EFFICIENCY AND RENEWABLE
     ENERGY

FROM: MARY ALICE HAYWARD
     SENIOR POLICY ADVISOR
     NATIONAL SECURITY MATTERS

SUBJECT: Nuclear Materials Disposition and Consolidation
     Coordination Committee

Former Secretary Abraham issued a memorandum (attached) on January 31, 2005, establishing the Nuclear Materials Disposition and Consolidation Coordination Committee. The principal mission of this Committee is to provide the forum to perform cross-cutting nuclear materials disposition and consolidation planning with an emphasis on providing the necessary security for our nuclear materials while reducing overall security costs and identifying paths for disposition, as appropriate. In my current position as Senior Policy Advisor on National Security Matters within the Department, I chair the Committee.

The Committee met for the first time today and, consistent with direction provided in memorandum, finalized a charter for your signature that fully documents the Committee's mission, scope, membership, meeting frequency, issue resolution procedures, and authorities. I have attached the charter for your review and approval.

I look forward to working with the Committee and developing and implementing a much needed Strategic Plan for the disposition and consolidation of the Department's nuclear materials assets. As Chairperson of this Committee, I will be reporting the Committee's progress to you on our efforts every six months.

Attachments

cc:

D. Garman, S-3	L. Brooks, NA-1
P. Golan, EM	E. Beckner, NA-10
W. Magwood, NE	P. Longsworth, NA-20
R. Orbach, SC	Admiral Kirk Donald, NA-30
G. Podonsky, SP	W. Desmond, NA-70

Printed with soy ink on recycled paper

Charter for the Nuclear Materials Disposition and Consolidation Coordination Committee (NMDCCC)

Mission

The principal mission of this Committee is to provide a forum to perform cross-cutting nuclear materials disposition and consolidation planning with the object of (1) providing the necessary security for our nuclear material, (2) identifying paths for disposition, as appropriate, and (3) reducing out-year security and program costs to the Department.

Scope

The Nuclear Materials Disposition and Consolidation Coordination Committee (NMDCCC) will address, coordinate, and take into account the Department's requirements for nuclear materials management, safeguards and security, and secure transportation and related issues as they pertain to nuclear materials disposition and consolidation and seek to leverage resources where practical. To that end, the NMDCCC will work to achieve the following:

- Develop, approve, revise, and ensure implementation of a Strategic Plan for disposition and consolidation of special nuclear material and an associated Implementation Schedule.
- Act, as necessary, to resolve conflicts created by priority use of departmental resources (secure transportation, packaging and containers, and waste management, etc) in concert with other established cross-cutting planning organizations, such as the Secure Transportation Asset Advisory Board, to assure that adverse impacts on program missions are minimized.
- Track and review progress against the approved Strategic Plan and Implementation Schedule and report on progress to the Secretary.

Membership

The NMDCCC operates under the direction of an Executive Steering Committee (ESC) comprised of the Under Secretary for Energy, Science, and Environment; the Under Secretary for Nuclear Security; and the Senior Policy Advisor for National Security Matters.

The NMDCCC is chaired by the Senior Policy Advisor for National Security Matters and its members are senior representatives of the Headquarters Program Offices with nuclear materials management, disposition, and safeguards and security responsibilities including the Offices of: Defense Programs; Secure Transportation; Defense Nuclear Nonproliferation; Naval Reactors; Defense Nuclear Security; Environmental Management; Nuclear Energy, Science, and Technology; Security and Safety Performance Assurance; Science; and Energy, Science and Environment (ESE) Security.

The NMDCCC will be supported, as needed, by ad hoc, task-oriented teams that may be drawn from the subject matter experts in the Headquarters programs, the Site Office staffs, and the management and operating contractor staffs. These ad hoc teams will be formed to address specific issues and serve at the pleasure of the ESC, dissolving when the issue has been addressed to the satisfaction of the ESC.

<u>Roles and Responsibilities</u>

<u>Executive Steering Committee</u>

The executive leaders are the champions for nuclear materials disposition and consolidation at the highest level within the Department. The executive leadership is empowered to establish priorities for resource allocation in coordination with the appropriate Program Secretarial Offices.

<u>Chairperson</u>

The Senior Policy Advisor for National Security Matters will chair the NMDCCC. The Chairperson is responsible for facilitating NMDCCC meetings, tasking the NMDCCC to conduct analyses and studies, and approving integrated nuclear materials disposition and consolidation schedules.

<u>NMDCCC Members</u>

NMDCCC members represent their respective organizations at NMDCCC meetings. Members provide recommendations to the Chairperson.

<u>Meetings</u>

The ESC will meet twice annually, or more frequently if necessary. The NMDCCC will meet more frequently in quarterly meetings and via monthly teleconference calls to assure effective, timely exchange of information and to promote coordination among programs and site representatives.

<u>Issue Resolution</u>

The normal channel for cross-cutting nuclear material disposition and consolidation issues to be identified and addressed by the ESC is through the NMDCCC. The NMDCCC is expected to try to resolve issues through coordination and consensus at that level and to keep the ESC informed. If an issue cannot be resolved at the NMDCCC level, it will be referred to ESC with explanation of the alternatives that the ESC must consider. If additional study of the issue is needed, the ESC may direct that a working group be formed by the NMDCCC to further study the problem and report its findings and recommendations to the ESC.

The NMDCCC Chairperson, in consultation with the NMDCCC members, will attempt to achieve a consensus on decisions related to nuclear material disposition and consolidation priorities and other related issues. The ESC will resolve any issues on which a consensus cannot be achieved

If consensus cannot be reached within the ESC on an issue, the Chairperson will raise that issue up to the Secretary for final disposition.

<u>Authority</u>

The NMDCCC does not in any way alter the line authority of the individual DOE/NNSA program organizations.

<u>Revisions</u>

The NMDCCC program prioritization guidelines will be administered by the NMDCCC and reviewed and revised as directed by the NMDCCC.

Approved:

Samuel W. Bodman
Secretary of Energy

March 4, 2005

Date

02:08:00 p.m.   10-03-2005   6/8

Department of Energy
Washington, DC 20585

February 15, 2005

MEMORANDUM FOR KEITH A. KLE[...]
MANAGER
RICHLAND O[...]

JEFFREY M. A[...]
MANAGER
SAVANNAH RIVER OPERATIONS OFFICE

FROM: PAUL M. GOLAN
                    ACTING ASSISTANT SECRETARY FOR
                    ENVIRONMENTAL MANAGEMENT

SUBJECT: Implementation Guidance for DOE Order 470.3 "Design
                    Basis Threat Policy"

On October 18, 2004, Deputy Secretary McSlarrow promulgated the Department of
Energy (DOE) Order 470.3 "Design Basis Threat Policy", which was sent to you
under secure cover. He is directing the preparation of new implementation plans by
July 29, 2005; accordingly, please submit your respective plans to me by July 11,
2005. The Deputy Secretary subsequently directed the implementation of
associated security upgrades to be completed by the end of fiscal year (FY) 2008
(attached).

In light of the new security policy, it is imperative that you continue to explore any
other ways to reduce your Category I targets or facilities through consolidation or
disposal. For example, I would like the Savannah River Operations Office (SR) to
provide me a recommendation concerning the feasibility of consolidating all
plutonium in the K Area Material Storage (KAMS) facility, rather than continuing
to upgrade 235-F.

In order to meet the Deputy Secretary's direction for new implementation plans by
next July, I believe that we should maximize our efficiencies by mobilizing all our
resources to achieve that goal. I recommend the suspension of any previously
scheduled activities that would conflict with the ultimate goal of implementing
DOE Order 470.3 by the end of FY08.

OPTIONAL FORM 99 (7-90)

FAX TRANSMITTAL   # of pages ▶ 2

To Ryan Coles   From Bob LaGrange

Dept./Agency   Phone #

Fax #   Fax #

NSN 7540-01-317-7368   5099-101   GENERAL SERVICES ADMINISTRATION

2

While it is our goal to consolidate and disposition EM owned nuclear materials as expeditiously as possible, some planning decisions still await senior Departmental approval.

For planning and analytical purposes, make the following additional assumptions:

- Hanford's Plutonium Finishing Plant (PFP) will be a non-enduring facility, but residual Category I materials may remain on the site through 2035. Specifically, some nuclear materials may be relocated from the PFP to another Hanford location. This location must provide protection of Category I materials against the 2004 DBT criteria.
- There will be only one enduring EM Category I facility at RL and one enduring EM Category I facility at SR.
- There will be no increased funds in FY05 or FY06. The implementation plans you submit in July will be used to support our out year budget requests and, as appropriate, any near term (FY06) budget reprogramming.

Please call me at (202) 586-7709 or Maurice Daugherty of my staff at (301) 903-9978 if you have any questions or need our assistance on these issues.

Attachment

cc:
D. German, HS-1
R. Wahb, S-3
G. Podonsky, SP-1
M. Comba, SO-1

APEX

TP_CF